Student Resources

INCLUDES

- Program Authors
- Table of Contents
- Picture Glossary
- Common Core State Standards Correlation
- Index

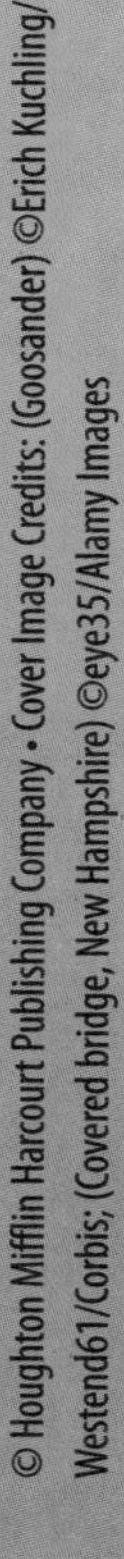

Made in the United States
Text printed on 100% recycled paper

Printed in the U.S.A.

ISBN 978-0-544-34345-0

19 0928 20

4500800118 B C D E F G

Dear Students and Families,

Welcome to **Go Math!**, Grade 2! In this exciting mathematics program, there are hands-on activities to do and real-world problems to solve. Best of all, you will write your ideas and answers right in your book. In **Go Math!**, writing and drawing on the pages helps you think deeply about what you are learning, and you will really understand math!

By the way, all of the pages in your **Go Math!** book are made using recycled paper. We wanted you to know that you can Go Green with **Go Math!**

Sincerely,

The Authors

Made in the United States
Text printed on 100% recycled paper

Authors

Juli K. Dixon, Ph.D.
Professor, Mathematics Education
University of Central Florida
Orlando, Florida

Edward B. Burger, Ph.D.
President, Southwestern University
Georgetown, Texas

Steven J. Leinwand
Principal Research Analyst
American Institutes for Research (AIR)
Washington, D.C.

Matthew R. Larson, Ph.D.
K-12 Curriculum Specialist for Mathematics
Lincoln Public Schools
Lincoln, Nebraska

Martha E. Sandoval-Martinez
Math Instructor
El Camino College
Torrance, California

Contributor

Rena Petrello
Professor, Mathematics
Moorpark College
Moorpark, CA

English Language Learners Consultant

Elizabeth Jiménez
CEO, GEMAS Consulting
Professional Expert on English Learner Education
Bilingual Education and Dual Language
Pomona, California

Table of Contents

Number Sense and Place Value

Critical Area

Critical Area Extending understanding of base-ten notation

Vocabulary Reader **Whales** 1

1 Number Concepts 9

Domains Operations and Algebraic Thinking
Number and Operations in Base Ten

COMMON CORE STATE STANDARDS 2.OA.C.3, 2.NBT.A.2, 2.NBT.A.3

✓ **Show What You Know** 10
Vocabulary Builder 11
Game: Three in a Row 12
1 **Hands On: Algebra • Even and Odd Numbers** 13
2 **Algebra • Represent Even Numbers** 19
3 **Understand Place Value** 25
4 **Expanded Form** 31
5 **Different Ways to Write Numbers** 37
✓ **Mid-Chapter Checkpoint** 40
6 **Algebra • Different Names for Numbers** 43
7 **Problem Solving • Tens and Ones** 49
8 **Counting Patterns Within 100** 55
9 **Counting Patterns Within 1,000** 61
✓ **Chapter 1 Review/Test** 67

GO DIGITAL

Go online! Your math lessons are interactive. Use *i*Tools, Animated Math Models, the Multimedia *e*Glossary, and more.

Chapter 1 Overview

In this chapter, you will explore and discover answers to the following **Essential Questions**:

- How do you use place value to find the values of numbers and describe numbers in different ways?
- How do you know the value of a digit?
- What are some different ways to show a number?
- How do you count by 1s, 5s, 10s, and 100s?

Chapter 2 Overview

In this chapter, you will explore and discover answers to the following **Essential Questions**:

- How can you use place value to model, write, and compare 3-digit numbers?
- How can you use blocks to show a 3-digit number?
- How can you write a 3-digit number in different ways?
- How can place value help you compare 3-digit numbers?

Practice and Homework

Lesson Check and Spiral Review in every lesson

Numbers to 1,000 71

Domain Number and Operations in Base Ten

COMMON CORE STATE STANDARDS 2.NBT.A.1, 2.NBT.A.1a, 2.NBT.A.1b, 2.NBT.A.3, 2.NBT.A.4, 2.NBT.B.8

✓ Show What You Know 72
Vocabulary Builder 73
Game: Fish for Digits 74
1 Group Tens as Hundreds 75
2 Explore 3-Digit Numbers 81
3 Hands On • Model 3-Digit Numbers 87
4 Hundreds, Tens, and Ones 93
5 Place Value to 1,000 99
6 Number Names 105
7 Different Forms of Numbers 111
✓ Mid-Chapter Checkpoint 114
8 Algebra • Different Ways to Show Numbers 117
9 Count On and Count Back by 10 and 100 123
10 Algebra • Number Patterns 129
11 Problem Solving • Compare Numbers 135
12 Algebra • Compare Numbers 141
✓ Chapter 2 Review/Test 147

Addition and Subtraction

Critical Area Building fluency with addition and subtraction

Vocabulary Reader All About Animals. 151

Basic Facts and Relationships 159

Domain Operations and Algebraic Thinking

COMMON CORE STATE STANDARDS 2.OA.A.1, 2.OA.B.2, 2.OA.C.4

✓ Show What You Know 160
Vocabulary Builder 161
Game: Caterpillar Chase 162
1 Use Doubles Facts. 163
2 Practice Addition Facts 169
3 Algebra • Make a Ten to Add. 175
4 Algebra • Add 3 Addends 181
5 Algebra • Relate Addition and Subtraction 187
6 Practice Subtraction Facts 193
✓ Mid-Chapter Checkpoint 196
7 Use Ten to Subtract 199
8 Algebra • Use Drawings to Represent Problems 205
9 Algebra • Use Equations to Represent Problems 211
10 Problem Solving • Equal Groups 217
11 Algebra • Repeated Addition. 223
✓ Chapter 3 Review/Test 229

Critical Area

GO DIGITAL

Go online! Your math lessons are interactive. Use *i*Tools, Animated Math Models, the Multimedia *e*Glossary, and more.

Chapter 3 Overview

In this chapter, you will explore and discover answers to the following **Essential Questions**:

- How can you use patterns and strategies to find sums and differences for basic facts?
- What are some strategies for remembering addition and subtraction facts?
- How are addition and subtraction related?

Chapter 4 Overview

In this chapter, you will explore and discover answers to the following **Essential Questions**:

- How do you use place value to add 2-digit numbers, and what are some different ways to add 2-digit numbers?
- How do you make an addend a ten to help solve an addition problem?
- How do you record the steps when adding 2-digit numbers?
- What are some ways to add 3 numbers or 4 numbers?

Practice and Homework

Lesson Check and Spiral Review in every lesson

2-Digit Addition 233

Domains Operations and Algebraic Thinking
Number and Operations in Base Ten

COMMON CORE STATE STANDARDS 2.OA.A.1, 2.NBT.B.5, 2.NBT.B.6, 2.NBT.B.9

✓ Show What You Know . . . 234
Vocabulary Builder . . . 235
Game: What is the Sum? . . . 236
1 Break Apart Ones to Add . . . 237
2 Use Compensation . . . 243
3 Break Apart Addends as Tens and Ones . . . 249
4 Model Regrouping for Addition . . . 255
5 Model and Record 2-Digit Addition . . . 261
6 2-Digit Addition . . . 267
7 Practice 2-Digit Addition . . . 273
✓ Mid-Chapter Checkpoint . . . 276
8 Rewrite 2-Digit Addition . . . 279
9 Problem Solving • Addition . . . 285
10 Algebra • Write Equations to Represent Addition . . . 291
11 Algebra • Find Sums for 3 Addends . . . 297
12 Algebra • Find Sums for 4 Addends . . . 303
✓ Chapter 4 Review/Test . . . 309

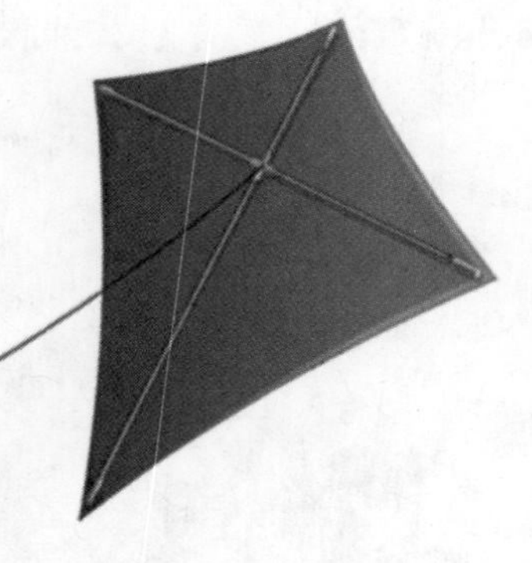

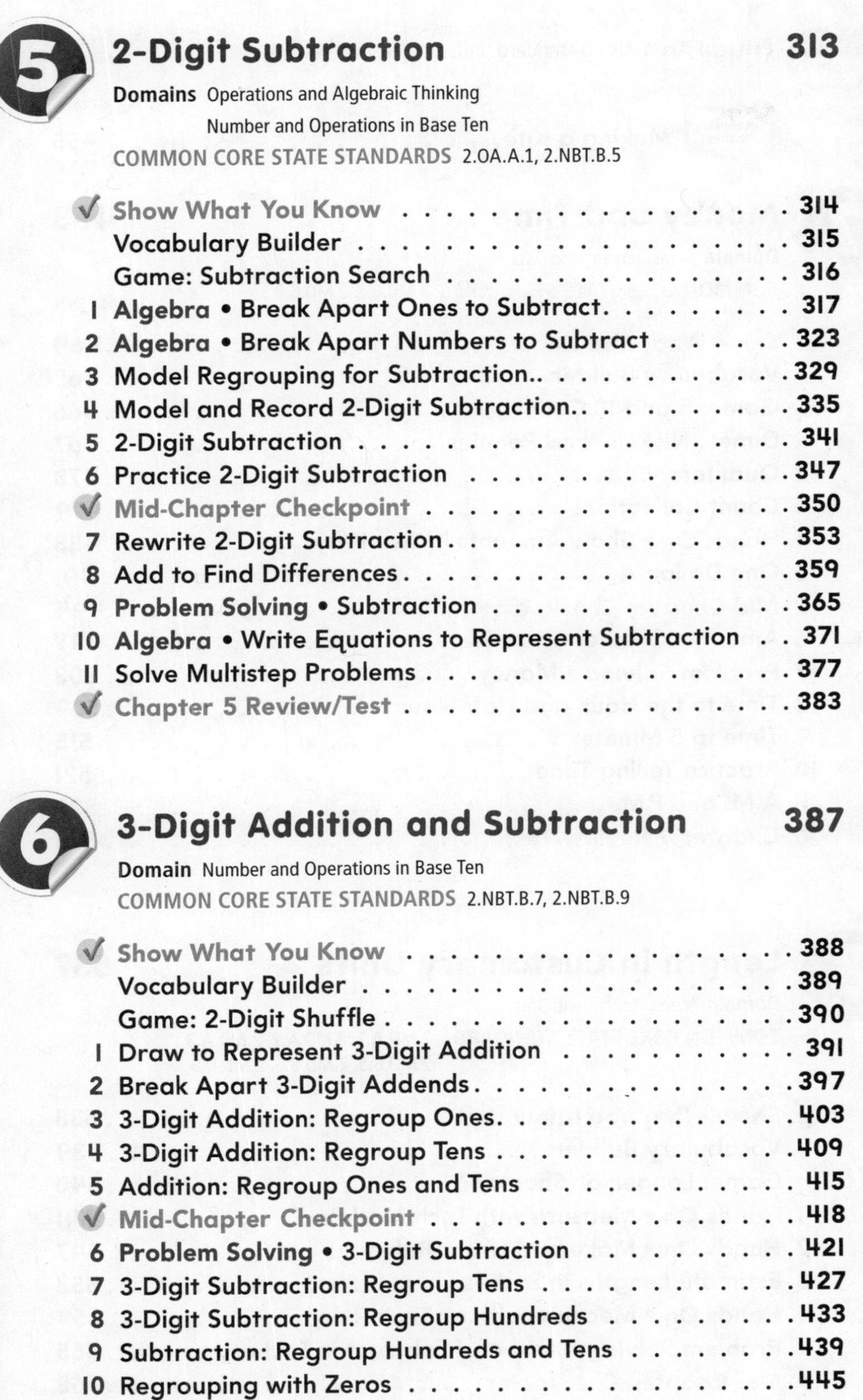

5 2-Digit Subtraction 313

Domains Operations and Algebraic Thinking
Number and Operations in Base Ten

COMMON CORE STATE STANDARDS 2.OA.A.1, 2.NBT.B.5

✓ Show What You Know 314
Vocabulary Builder 315
Game: Subtraction Search 316
1 Algebra • Break Apart Ones to Subtract 317
2 Algebra • Break Apart Numbers to Subtract 323
3 Model Regrouping for Subtraction 329
4 Model and Record 2-Digit Subtraction 335
5 2-Digit Subtraction 341
6 Practice 2-Digit Subtraction 347
✓ Mid-Chapter Checkpoint 350
7 Rewrite 2-Digit Subtraction 353
8 Add to Find Differences 359
9 Problem Solving • Subtraction 365
10 Algebra • Write Equations to Represent Subtraction . . 371
11 Solve Multistep Problems 377
✓ Chapter 5 Review/Test 383

Chapter 5 Overview

In this chapter, you will explore and discover answers to the following **Essential Questions**:

- How do you use place value to subtract 2-digit numbers with and without regrouping?
- How can you break apart numbers to help solve a subtraction problem?
- What are the steps you use when you solve 2-digit subtraction problems?
- What are some different ways to model, show, and solve subtraction problems?

6 3-Digit Addition and Subtraction 387

Domain Number and Operations in Base Ten

COMMON CORE STATE STANDARDS 2.NBT.B.7, 2.NBT.B.9

✓ Show What You Know 388
Vocabulary Builder 389
Game: 2-Digit Shuffle 390
1 Draw to Represent 3-Digit Addition 391
2 Break Apart 3-Digit Addends 397
3 3-Digit Addition: Regroup Ones 403
4 3-Digit Addition: Regroup Tens 409
5 Addition: Regroup Ones and Tens 415
✓ Mid-Chapter Checkpoint 418
6 Problem Solving • 3-Digit Subtraction 421
7 3-Digit Subtraction: Regroup Tens 427
8 3-Digit Subtraction: Regroup Hundreds 433
9 Subtraction: Regroup Hundreds and Tens 439
10 Regrouping with Zeros 445
✓ Chapter 6 Review/Test 451

Chapter 6 Overview

In this chapter, you will explore and discover answers to the following **Essential Questions**:

- What are some strategies for adding and subtracting 3-digit numbers?
- What are the steps when finding the sum in a 3-digit addition problem?
- What are the steps when finding the difference in a 3-digit subtraction problem?
- When do you need to regroup?

Critical Area

GO DIGITAL

Go online! Your math lessons are interactive. Use *i*Tools, Animated Math Models, the Multimedia *e*Glossary, and more.

Chapter 7 Overview

Essential Questions:
- How do you use the values of coins and bills to find the total value of a group of money, and how do you read times shown on analog and digital clocks?
- What are the names and values of the different coins?
- How can you tell the time on a clock by looking at the clock hands?

Chapter 8 Overview

Essential Questions:
- What are some of the methods and tools that can be used to estimate and measure length?
- What tools can be used to measure length and how do you use them?
- What units can be used to measure length and how do they compare with each other?
- How can you estimate the length of an object?

Measurement and Data

Common Core **Critical Area** Using standard units of measure

Vocabulary Reader Making a Kite 455

7 Money and Time 463

Domain Measurement and Data
COMMON CORE STATE STANDARDS 2.MD.C.7, 2.MD.C.8

Show What You Know 464
Vocabulary Builder 465
Game: 5 and 10 Count 466
1 Dimes, Nickels, and Pennies 467
2 Quarters 473
3 Count Collections 479
4 Hands On • Show Amounts in Two Ways 485
5 One Dollar 491
Mid-Chapter Checkpoint 494
6 Amounts Greater Than $1 497
7 Problem Solving • Money 503
8 Time to the Hour and Half Hour 509
9 Time to 5 Minutes 515
10 Practice Telling Time 521
11 A.M. and P.M. 527
Chapter 7 Review/Test 533

8 Length in Customary Units 537

Domain Measurement and Data
COMMON CORE STATE STANDARDS 2.MD.A.1, 2.MD.A.2, 2.MD.A.3, 2.MD.B.5, 2.MD.B.6, 2.MD.D.9

Show What You Know 538
Vocabulary Builder 539
Game: Longer or Shorter? 540
1 Hands On • Measure with Inch Models 541
2 Hands On • Make and Use a Ruler 547
3 Estimate Lengths in Inches 553
4 Hands On • Measure with an Inch Ruler 559
5 Problem Solving • Add and Subtract in Inches 565
Mid-Chapter Checkpoint 568
6 Hands On • Measure in Inches and Feet 571
7 Estimate Lengths in Feet 577
8 Choose a Tool 583
9 Display Measurement Data 589
Chapter 8 Review/Test 595

9 Length in Metric Units 599

Domain Measurement and Data

COMMON CORE STATE STANDARDS 2.MD.A.1, 2.MD.A.2, 2.MD.A.3, 2.MD.A.4, 2.MD.B.5, 2.MD.B.6

✓ Show What You Know 600
Vocabulary Builder 601
Game: Estimating Length 602
1 Hands On • Measure with a Centimeter Model 603
2 Estimate Lengths in Centimeters 609
3 Hands On • Measure with a Centimeter Ruler 615
4 Problem Solving • Add and Subtract Lengths 621
✓ Mid-Chapter Checkpoint 624
5 Hands On • Centimeters and Meters 627
6 Estimate Lengths in Meters 633
7 Hands On • Measure and Compare Lengths 639
✓ Chapter 9 Review/Test 645

10 Data 649

Domain Measurement and Data

COMMON CORE STATE STANDARDS 2.MD.D.10

✓ Show What You Know 650
Vocabulary Builder 651
Game: Making Tens 652
1 Collect Data 653
2 Read Picture Graphs 659
3 Make Picture Graphs 665
✓ Mid-Chapter Checkpoint 668
4 Read Bar Graphs 671
5 Make Bar Graphs 677
6 Problem Solving • Display Data 683
✓ Chapter 10 Review/Test 689

Chapter 9 Overview

In this chapter, you will explore and discover answers to the following **Essential Questions**:

- What are some of the methods and tools that can be used to estimate and measure length in metric units?
- What tools can be used to measure length in metric units and how do you use them?
- What metric units can be used to measure length and how do they compare with each other?
- If you know the length of one object, how can you estimate the length of another object?

Practice and Homework

Lesson Check and Spiral Review in every lesson

Chapter 10 Overview

In this chapter, you will explore and discover answers to the following **Essential Questions**:

- How do tally charts, picture graphs, and bar graphs help you solve problems?
- How are tally marks used to record data for a survey?
- How is a picture graph made?
- How do you know what the bars in a bar graph stand for?

Critical Area

Geometry and Fractions

Critical Area Describing and analyzing shapes

Vocabulary Reader A Farmer's Job. 693

11 Geometry and Fraction Concepts 701

Domain Geometry

COMMON CORE STATE STANDARDS 2.G.A.1, 2.G.A.2, 2.G.A.3

- ✓ Show What You Know 702
- Vocabulary Builder 703
- Game: Count the Sides 704
- 1 Three-Dimensional Shapes 705
- 2 Attributes of Three-Dimensional Shapes 711
- 3 Build Three-Dimensional Shapes 717
- 4 Two-Dimensional Shapes 723
- 5 Angles in Two-Dimensional Shapes 729
- 6 Sort Two-Dimensional Shapes. 735
- 7 Hands On • Partition Rectangles 741
- ✓ Mid-Chapter Checkpoint 744
- 8 Equal Parts 747
- 9 Show Equal Parts of a Whole. 753
- 10 Describe Equal Parts 759
- 11 Problem Solving • Equal Shares 765
- ✓ Chapter 11 Review/Test 771

Picture Glossary H1
Common Core State Standards H11
Index H17

GO DIGITAL

Go online! Your math lessons are interactive. Use *i*Tools, Animated Math Models, the Multimedia *e*Glossary, and more.

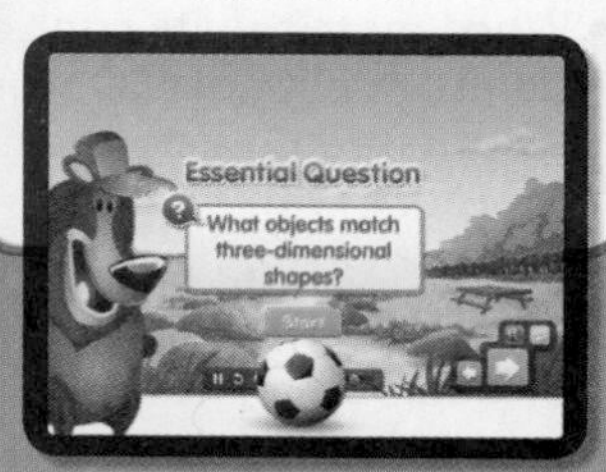

Chapter 11 Overview

In this chapter, you will explore and discover answers to the following **Essential Questions**:

- What are some two-dimensional shapes and three-dimensional shapes, and how can you show equal parts of shapes?
- How can you describe some two-dimensional and three-dimensional shapes?
- How can you describe equal parts of shapes?

Picture Glossary

addend sumando

5 + 8 = 13

addends

bar graph gráfica de barras

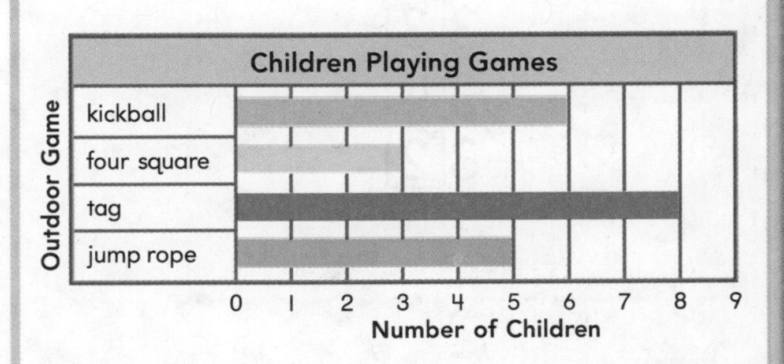

a.m. a.m.

Times after midnight and before noon are written with **a.m.**

11:00 a.m. is in the morning.

cent sign símbolo de centavo

53¢

cent sign

angle ángulo

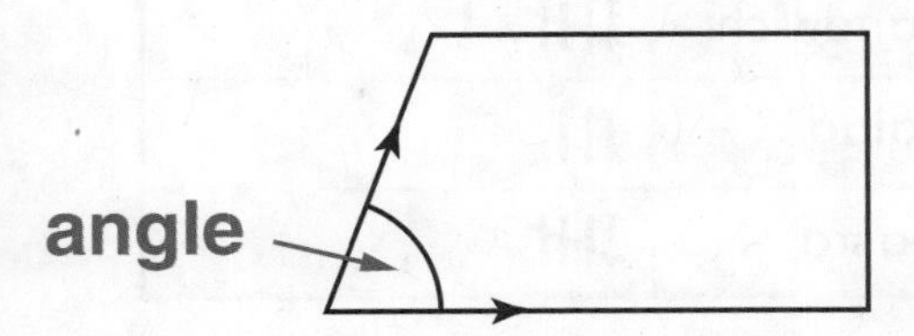

centimeter centímetro

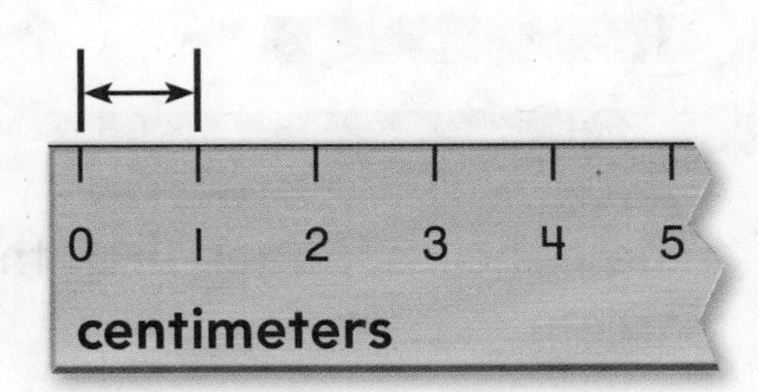

column columna

column

$$\begin{array}{r} 3\boxed{3} \\ 3\boxed{4} \\ +3\boxed{2} \\ \hline \end{array}$$

compare comparar

Use these symbols when you **compare**: >, <, =.

241 > 234

123 < 128

247 = 247

compare comparar

Compare the lengths of the pencil and the crayon.

The pencil is longer than the crayon.

The crayon is shorter than the pencil.

cone cono

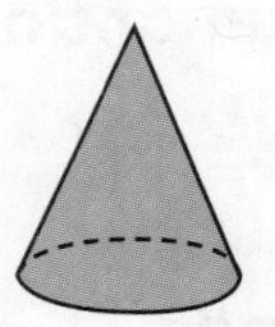

cube cubo

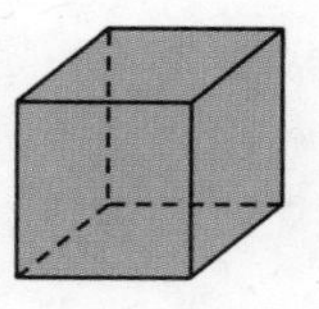

cylinder cilindro

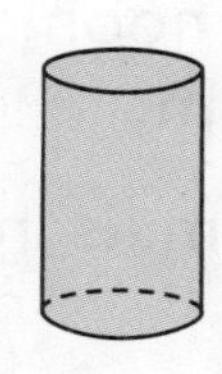

data datos

Favorite Lunch	
Lunch	**Tally**
pizza	IIII
sandwich	~~IIII~~ I
salad	III
pasta	~~IIII~~

The information in this chart is called **data**.

decimal point punto decimal

$1.00
↑
decimal point

difference diferencia

9 − 2 = 7
↑
difference

digit dígito

0, 1, 2, 3, 4, 5, 6, 7, 8, and 9 are **digits**.

dime moneda de 10¢

A **dime** has a value of 10 cents.

dollar dólar

One **dollar** is worth 100 cents.

dollar sign símbolo de dólar

$1.00
↑
dollar sign

doubles dobles

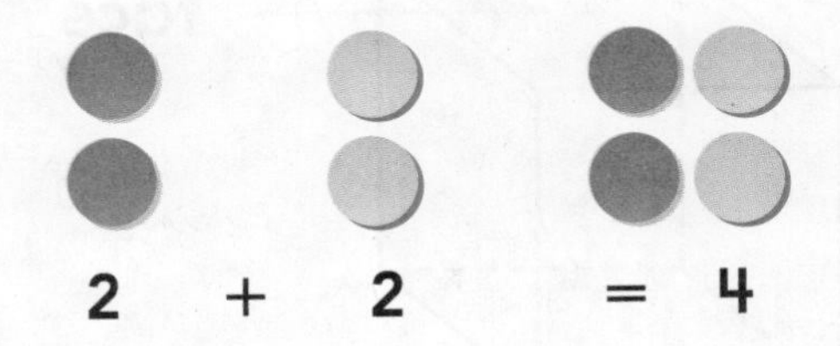

edge arista

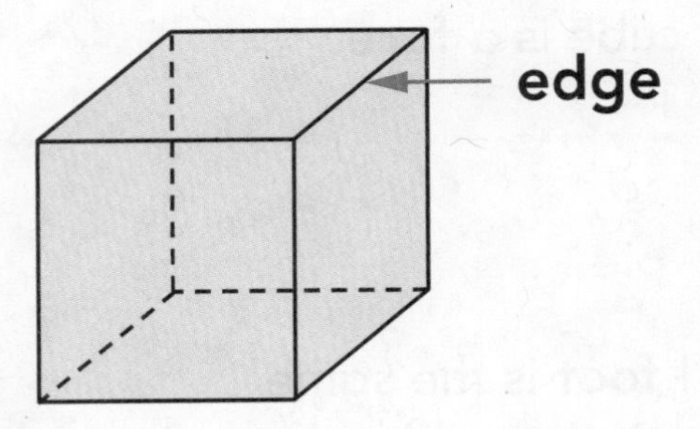

An **edge** is formed where two faces of a three-dimensional shape meet.

estimate estimación

An **estimate** is an amount that tells about how many.

even par

2, 4, 6, 8, 10, . . .

even numbers

face cara

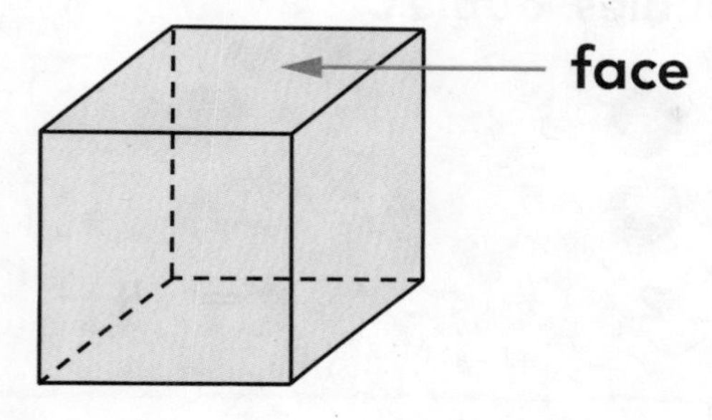

Each flat surface of this cube is a **face**.

foot pie

1 **foot** is the same length as 12 inches.

fourth of cuarto de

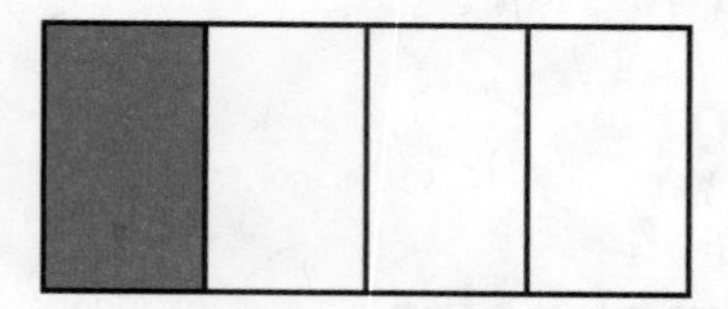

A **fourth of** the shape is green.

fourths cuartos

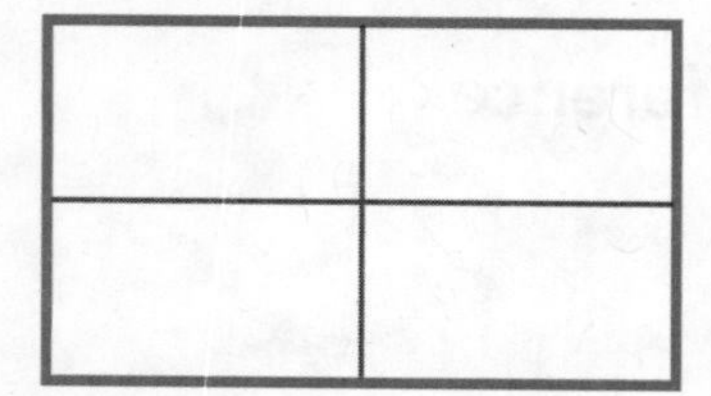

This shape has 4 equal parts. These equal parts are called **fourths**.

half of mitad de

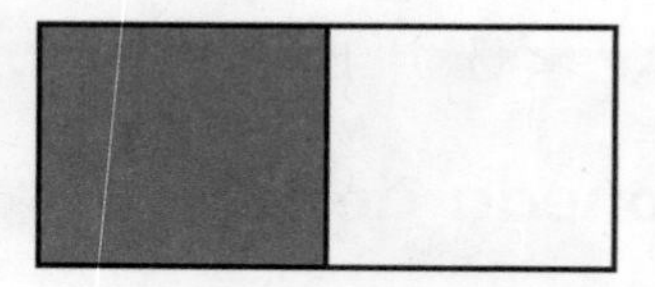

A **half of** the shape is green.

halves mitades

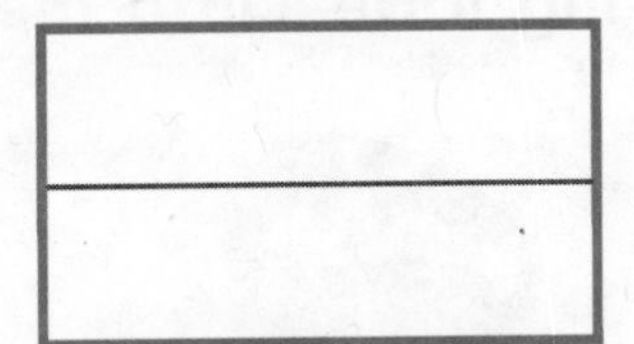

This shape has 2 equal parts. These equal parts are called **halves**.

hexagon hexágono

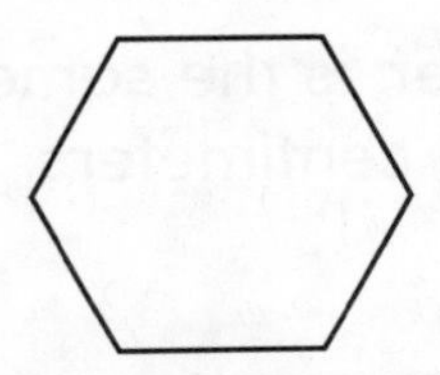

A two-dimensional shape with 6 sides is a **hexagon**.

hour hora

There are 60 minutes in 1 **hour**.

hundred centena

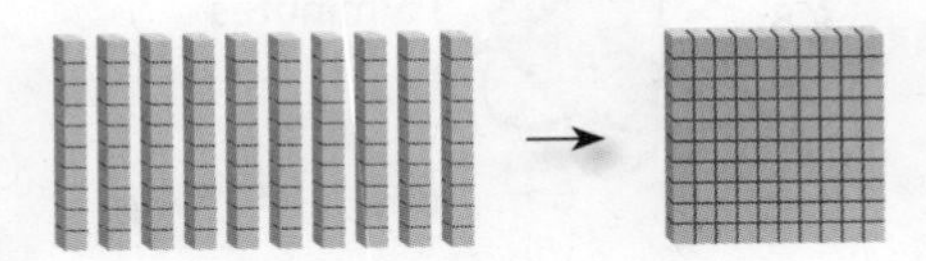

There are 10 tens in 1 **hundred**.

inch pulgada

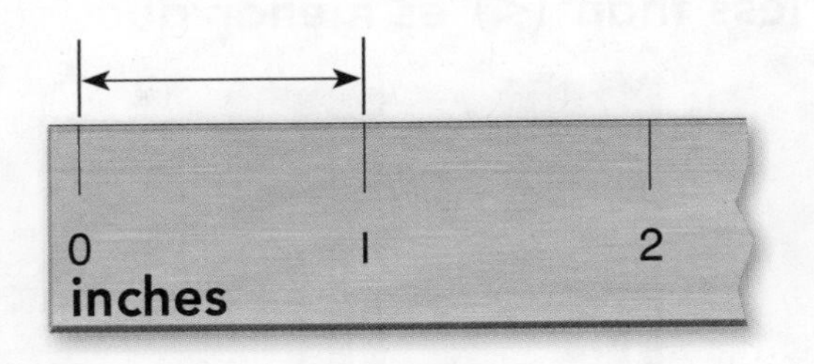

is equal to (=) es igual a

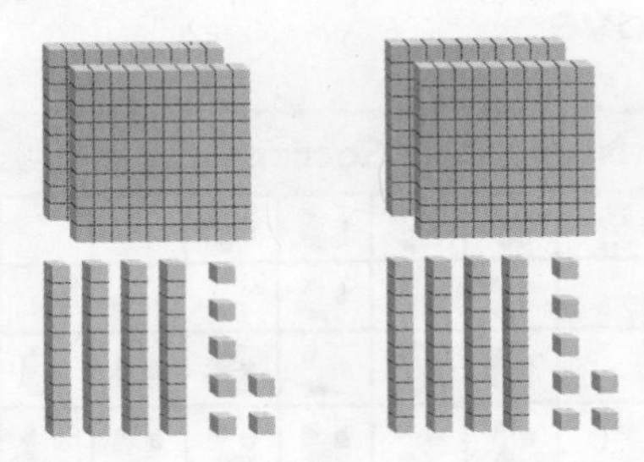

247 **is equal to** 247.

247 = 247

is greater than (>) es mayor que

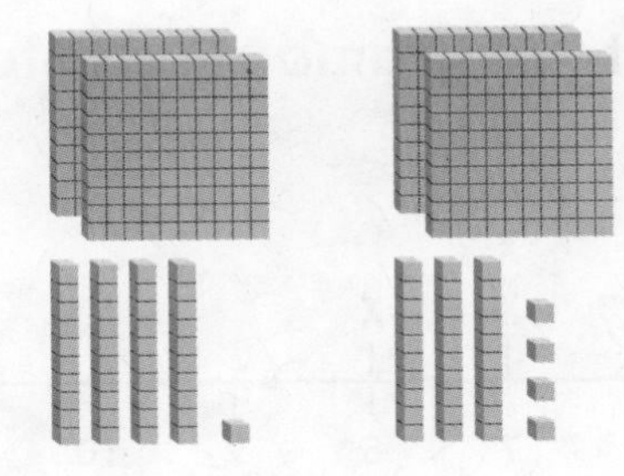

241 **is greater than** 234.

241 > 234

is less than (<) es menor que

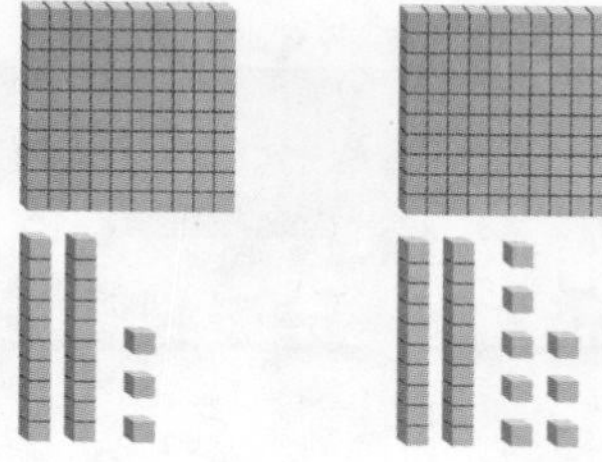

123 **is less than** 128.

$123 < 128$

---

measuring tape cinta métrica

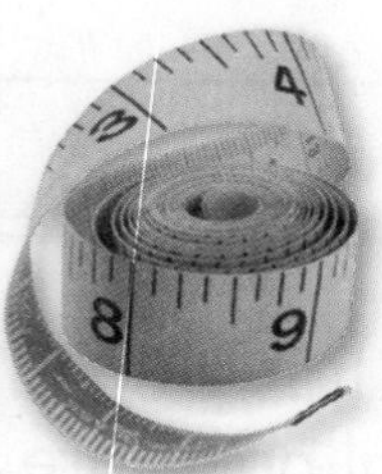

---

key clave

Number of Soccer Games							
March	⚽	⚽	⚽	⚽			
April	⚽	⚽	⚽				
May	⚽	⚽	⚽	⚽	⚽	⚽	
June	⚽	⚽	⚽	⚽	⚽	⚽	⚽

Key: Each ⚽ stands for 1 game.

The **key** tells how many each picture stands for.

---

meter metro

1 **meter** is the same length as 100 centimeters.

---

midnight medianoche

Midnight is 12:00 at night.

---

line plot diagrama de puntos

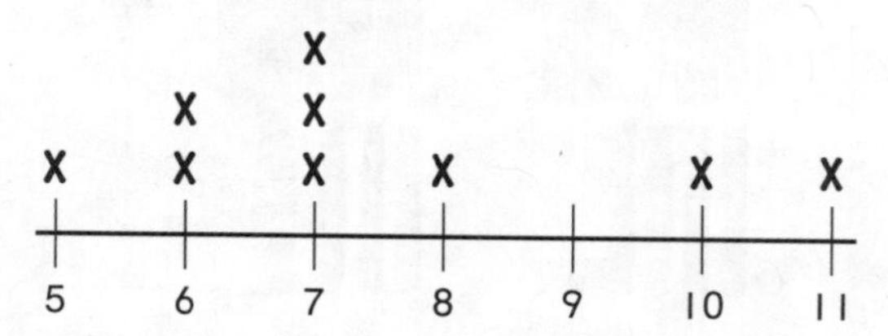

Lengths of Paintbrushes in Inches

---

minute minuto

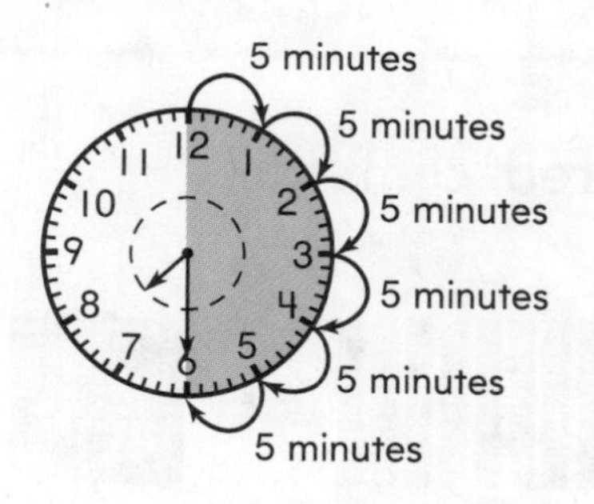

There are 30 **minutes** in a half hour.

nickel moneda de 5¢

A **nickel** has a value of 5 cents.

noon mediodía

Noon is 12:00 in the daytime.

odd impar

1, 3, 5, 7, 9, 11, . . .

odd numbers

ones unidades

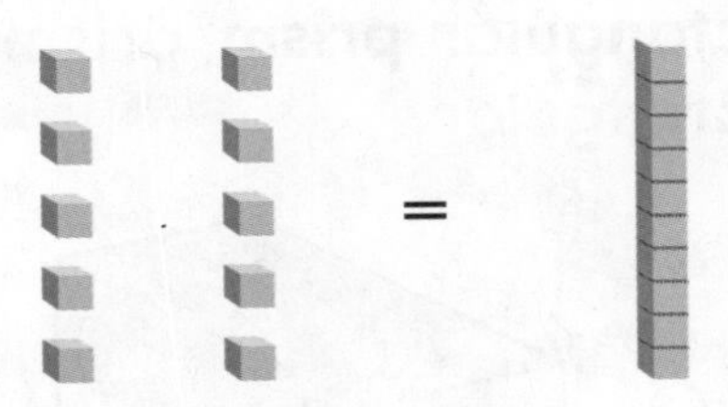

10 ones = 1 ten

penny moneda de 1¢

A **penny** has a value of 1 cent.

pentagon pentágono

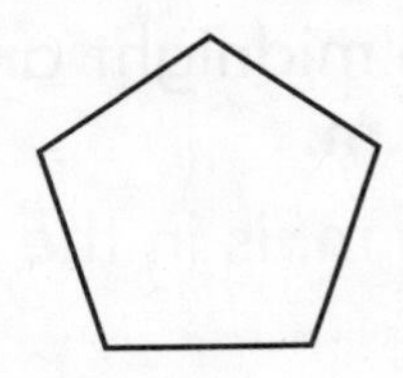

A two-dimensional shape with 5 sides is a **pentagon**.

picture graph gráfica con dibujos

Number of Soccer Games							
March	⚽	⚽	⚽	⚽			
April	⚽	⚽	⚽				
May	⚽	⚽	⚽	⚽	⚽	⚽	
June	⚽	⚽	⚽	⚽	⚽	⚽	⚽

Key: Each ⚽ stands for 1 game.

plus (+) más

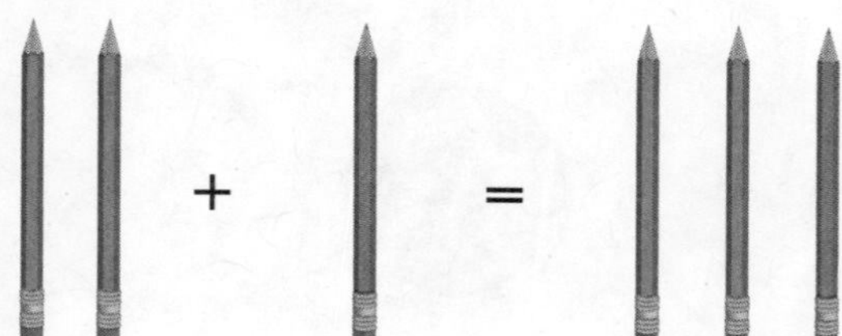

2 plus 1 is equal to 3

2 + 1 = 3

p.m. p.m.

Times after noon and before midnight are written with **p.m.**

11:00 p.m. is in the evening.

quadrilateral cuadrilátero

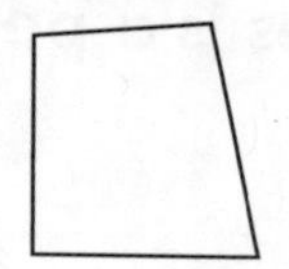

A two-dimensional shape with 4 sides is a **quadrilateral**.

quarter moneda de 25¢

A **quarter** has a value of 25 cents.

quarter of cuarta parte de

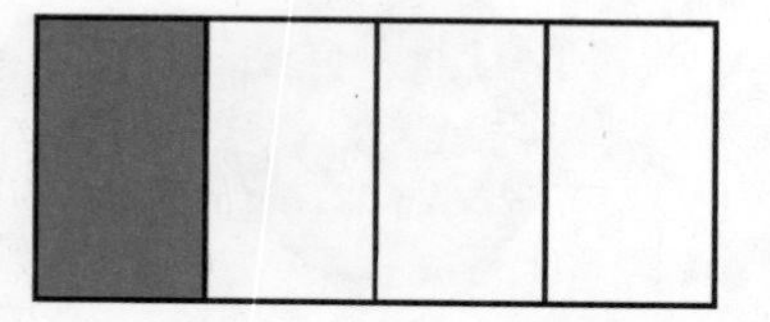

A **quarter of** the shape is green.

quarter past y cuarto

15 minutes after 8
quarter past 8

rectangular prism prisma rectangular

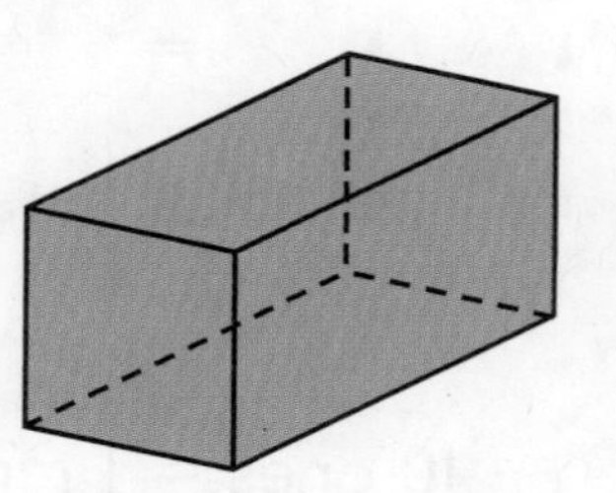

regroup reagrupar

Tens | Ones

You can trade 10 ones for 1 ten to **regroup**.

side lado

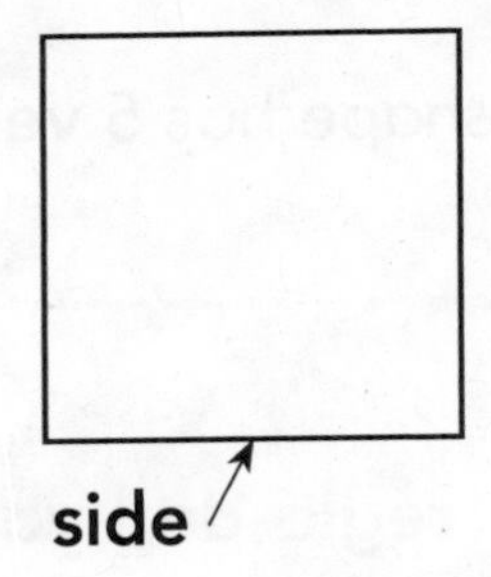

This shape has 4 **sides**.

sphere esfera

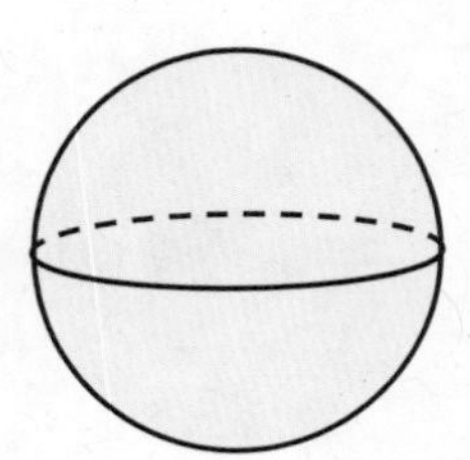

sum suma o total

$9 + 6 = 15$

sum

survey encuesta

Favorite Lunch	
Lunch	**Tally**
pizza	IIII
sandwich	~~IIII~~ I
salad	III
pasta	~~IIII~~

A **survey** is a collection of data from answers to a question.

ten decena

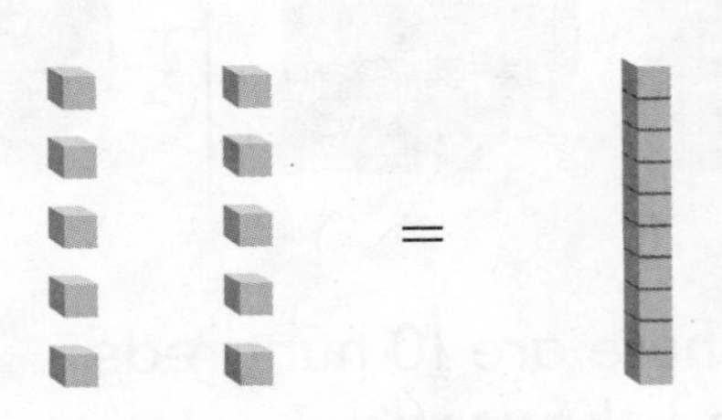

10 ones = 1 ten

third of tercio de

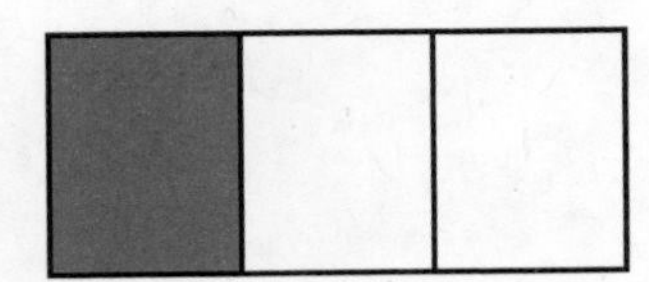

A **third of** the shape is green.

thirds tercios

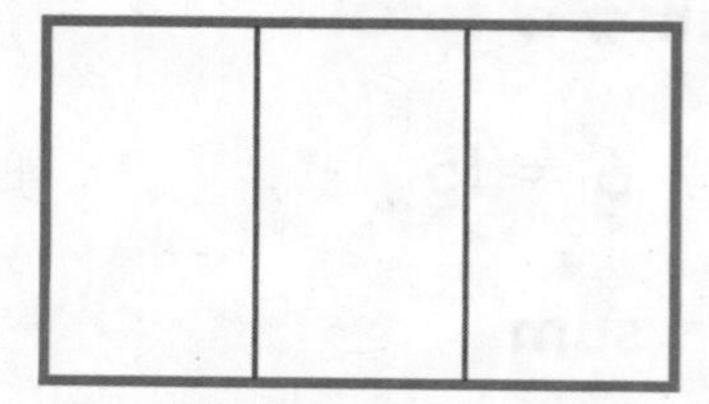

This shape has 3 equal parts. These equal parts are called **thirds**.

thousand millar

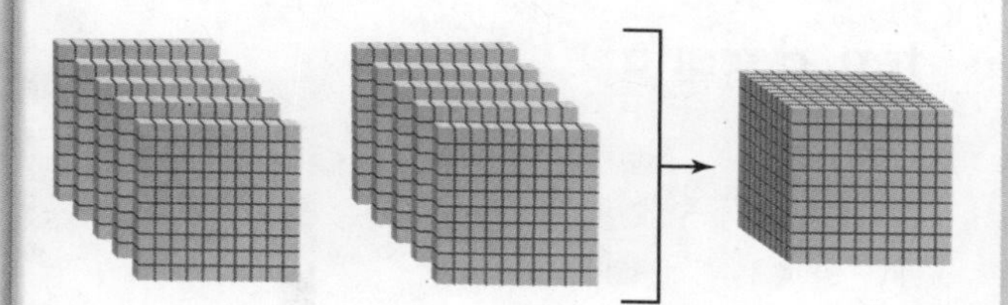

There are 10 hundreds in 1 **thousand**.

vertex/vertices vértice/vértices

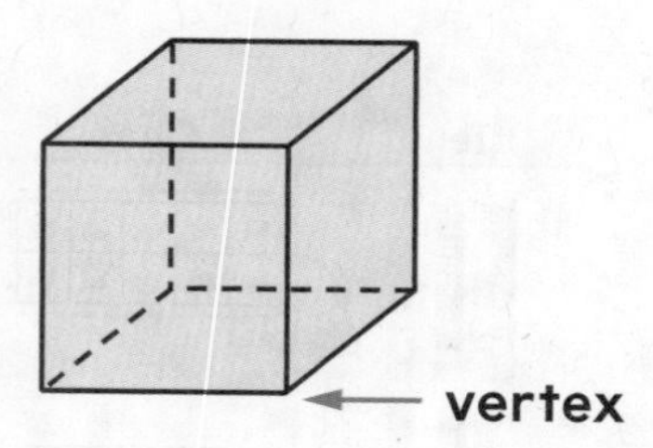

A corner point of a three-dimensional shape is a **vertex**.

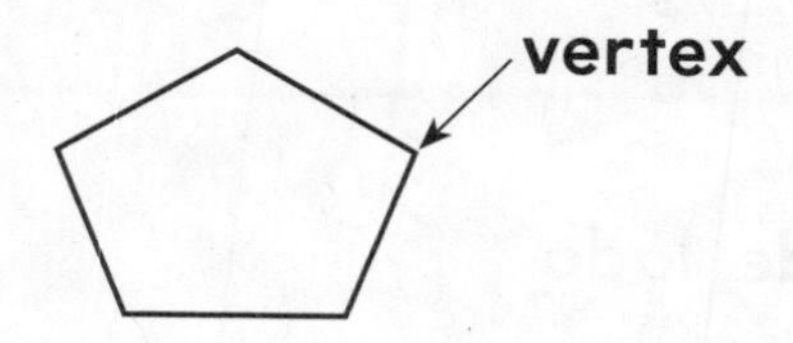

This shape has 5 **vertices**.

yardstick regla de 1 yarda

A **yardstick** is a measuring tool that shows 3 feet.

Correlations

COMMON CORE STATE STANDARDS

Standards You Will Learn

Mathematical Practices		Some examples are:
MP1	Make sense of problems and persevere in solving them.	Lessons 1.3, 1.5, 2.2, 3.2, 3.3, 4.7, 4.9, 4.11, 5.9, 5.10, 5.11, 6.7, 7.7, 8.5, 9.4, 10.1, 10.2, 10.3, 10.4, 10.6, 11.5
MP2	Reason abstractly and quantitatively.	Lessons 1.2, 2.6, 2.11, 2.12, 3.5, 3.9, 4.9, 4.10, 5.5, 5.9, 5.10, 5.11, 6.1, 8.1, 8.4, 8.5, 8.6, 9.4, 9.7, 10.2, 10.4
MP3	Construct viable arguments and critique the reasoning of others.	Lessons 1.1, 2.5, 2.8, 2.11, 4.6, 5.6, 5.10, 6.8, 8.8, 9.3, 10.5, 10.6
MP4	Model with mathematics.	Lessons 1.4, 1.7, 2.3, 2.11, 3.8, 3.9, 3.11, 4.1, 4.2, 4.5, 4.9, 4.11, 5.4, 5.9, 5.10, 5.11, 6.6, 7.3, 7.4, 7.5, 7.6, 7.7, 8.5, 8.9, 9.4, 10.1, 10.3, 10.5, 10.6, 11.3, 11.4, 11.5, 11.6, 11.10, 11.11
MP5	Use appropriate tools strategically.	Lessons 3.7, 3.10, 4.4, 5.1, 5.2, 5.3, 5.8, 6.1, 8.1, 8.2, 8.4, 8.6, 8.8, 8.9, 9.1, 9.3, 9.5, 11.2, 11.7, 11.9
MP6	Attend to precision.	Lessons 1.3, 1.5, 1.6, 2.1, 2.5, 2.7, 2.12, 3.5, 3.11, 4.1, 4.2, 4.3, 4.6, 4.8, 4.10, 4.12, 5.5, 5.7, 6.1, 6.2, 6.3, 6.4, 6.5, 6.7, 6.9, 6.10, 7.2, 7.3, 7.8, 7.9, 7.10, 7.11, 8.1, 8.2, 8.3, 8.4, 8.6, 8.7, 8.9, 9.1, 9.2, 9.3, 9.6, 9.7, 10.1, 10.2, 10.3, 10.4, 10.5, 11.1, 11.2, 11.6, 11.8, 11.9, 11.10, 11.11
MP7	Look for and make use of structure.	Lessons 1.1, 1.2, 1.6, 1.7, 1.8, 1.9, 2.1, 2.2, 2.3, 2.4, 2.5, 2.6, 2.7, 2.8, 2.9, 2.10, 3.1, 3.2, 3.3, 3.7, 3.10, 4.4, 4.7, 4.8, 5.3, 5.6, 5.7, 7.1, 7.5, 7.6, 7.11, 8.3, 8.7, 9.2, 9.5, 9.6, 11.4, 11.5

Standards You Will Learn

Mathematical Practices		Some examples are:
MP8	Look for and express regularity in repeated reasoning.	Lessons 1.2, 1.6, 2.1, 2.2, 2.4, 2.12, 3.2, 3.3, 3.4, 3.5, 3.7, 4.3, 4.11, 4.12, 5.5, 5.8, 6.2, 6.3, 6.4, 6.5, 6.7, 6.8, 6.9, 6.10, 7.2, 7.3, 7.4, 7.8, 7.9, 7.10, 8.1, 8.8, 9.1, 11.7, 11.8
Domain: Operations and Algebraic Thinking		**Student Edition Lessons**
Represent and solve problems involving addition and subtraction.		
2.OA.A.1	Use addition and subtraction within 100 to solve one- and two-step word problems involving situations of adding to, taking from, putting together, taking apart, and comparing, with unknowns in all positions, e.g., by using drawings and equations with a symbol for the unknown number to represent the problem.	Lessons 3.8, 3.9, 4.9, 4.10, 5.9, 5.10, 5.11
Add and subtract within 20.		
2.OA.B.2	Fluently add and subtract within 20 using mental strategies. By end of Grade 2, know from memory all sums of two one-digit numbers.	Lessons 3.1, 3.2, 3.3, 3.4, 3.5, 3.6, 3.7
Work with equal groups of objects to gain foundations for multiplication.		
2.OA.C.3	Determine whether a group of objects (up to 20) has an odd or even number of members, e.g., by pairing objects or counting them by 2s; write an equation to express an even number as a sum of two equal addends.	Lessons 1.1, 1.2
2.OA.C.4	Use addition to find the total number of objects arranged in rectangular arrays with up to 5 rows and up to 5 columns; write an equation to express the total as a sum of equal addends.	Lessons 3.10, 3.11

Standards You Will Learn

Standards You Will Learn		Student Edition Lessons
Domain: Number and Operations in Base Ten		
Understand place value.		
2.NBT.A.1	Understand that the three digits of a three-digit number represent amounts of hundreds, tens, and ones; e.g., 706 equals 7 hundreds, 0 tens, and 6 ones. Understand the following as special cases:	Lessons 2.2, 2.3, 2.4, 2.5
	a. 100 can be thought of as a bundle of ten tens — called a "hundred."	Lesson 2.1
	b. The numbers 100, 200, 300, 400, 500, 600, 700, 800, 900 refer to one, two, three, four, five, six, seven, eight, or nine hundreds (and 0 tens and 0 ones).	Lesson 2.1
2.NBT.A.2	Count within 1000; skip-count by 5s, 10s, and 100s.	Lessons 1.8, 1.9
2.NBT.A.3	Read and write numbers to 1000 using base-ten numerals, number names, and expanded form.	Lessons 1.3, 1.4, 1.5. 1.6, 1.7, 2.4, 2.6, 2.7, 2.8
2.NBT.A.4	Compare two three-digit numbers based on meanings of the hundreds, tens, and ones digits, using >, =, and < symbols to record the results of comparisons.	Lessons 2.11, 2.12
Use place value understanding and properties of operations to add and subtract.		
2.NBT.B.5	Fluently add and subtract within 100 using strategies based on place value, properties of operations, and/or the relationship between addition and subtraction.	Lessons 4.1, 4.2, 4.3, 4.4, 4.5, 4.6, 4.7, 4.8, 5.1, 5.2, 5.3, 5.4, 5.5, 5.6, 5.7, 5.8
2.NBT.B.6	Add up to four two-digit numbers using strategies based on place value and properties of operations.	Lessons 4.11, 4.12

Standards You Will Learn

	Standards You Will Learn	Student Edition Lessons
Domain: Number and Operations in Base Ten		
Use place value understanding and properties of operations to add and subtract.		
2.NBT.B.7	Add and subtract within 1000, using concrete models or drawings and strategies based on place value, properties of operations, and/or the relationship between addition and subtraction; relate the strategy to a written method. Understand that in adding or subtracting three-digit numbers, one adds or subtracts hundreds and hundreds, tens and tens, ones and ones; and sometimes it is necessary to compose or decompose tens or hundreds.	Lessons 6.1, 6.2, 6.3, 6.4, 6.5, 6.6, 6.7, 6.8, 6.9, 6.10
2.NBT.B.8	Mentally add 10 or 100 to a given number 100–900, and mentally subtract 10 or 100 from a given number 100–900.	Lessons 2.9, 2.10
2.NBT.B.9	Explain why addition and subtraction strategies work, using place value and the properties of operations.	Lessons 4.6, 6.8
Domain: Measurement and Data		
Measure and estimate lengths in standard units.		
2.MD.A.1	Measure the length of an object by selecting and using appropriate tools such as rulers, yardsticks, meter sticks, and measuring tapes.	Lessons 8.1, 8.2, 8.4, 8.8, 9.1, 9.3
2.MD.A.2	Measure the length of an object twice, using length units of different lengths for the two measurements; describe how the two measurements relate to the size of the unit chosen.	Lessons 8.6, 9.5

Standards You Will Learn

		Student Edition Lessons
Domain: Measurement and Data		
Measure and estimate lengths in standard units.		
2.MD.A.3	Estimate lengths using units of inches, feet, centimeters, and meters.	Lessons 8.3, 8.7, 9.2, 9.6
2.MD.A.4	Measure to determine how much longer one object is than another, expressing the length difference in terms of a standard length unit.	Lesson 9.7
Relate addition and subtraction to length.		
2.MD.B.5	Use addition and subtraction within 100 to solve word problems involving lengths that are given in the same units, e.g., by using drawings (such as drawings of rulers) and equations with a symbol for the unknown number to represent the problem.	Lessons 8.5, 9.4
2.MD.B.6	Represent whole numbers as lengths from 0 on a number line diagram with equally spaced points corresponding to the numbers 0, 1, 2, ..., and represent whole-number sums and differences within 100 on a number line diagram.	Lessons 8.5, 9.4
Work with time and money.		
2.MD.C.7	Tell and write time from analog and digital clocks to the nearest five minutes, using a.m. and p.m.	Lessons 7.8, 7.9, 7.10, 7.11
2.MD.C.8	Solve word problems involving dollar bills, quarters, dimes, nickels, and pennies, using $ and ¢ symbols appropriately. *Example: If you have 2 dimes and 3 pennies, how many cents do you have?*	Lessons 7.1, 7.2, 7.3, 7.4, 7.5, 7.6, 7.7

Standards You Will Learn

	Standards You Will Learn	Student Edition Lessons
Domain: Measurement and Data		
Represent and interpret data.		
2.MD.D.9	Generate measurement data by measuring lengths of several objects to the nearest whole unit, or by making repeated measurements of the same object. Show the measurements by making a line plot, where the horizontal scale is marked off in whole-number units.	Lesson 8.9
2.MD.D.10	Draw a picture graph and a bar graph (with single-unit scale) to represent a data set with up to four categories. Solve simple put-together, take-apart, and compare problems using information presented in a bar graph.	Lessons 10.1, 10.2, 10.3, 10.4, 10.5, 10.6
Domain: Geometry		
Reason with shapes and their attributes.		
2.G.A.1	Recognize and draw shapes having specified attributes, such as a given number of angles or a given number of equal faces. Identify triangles, quadrilaterals, pentagons, hexagons, and cubes.	Lessons 11.1, 11.2, 11.3, 11.4, 11.5, 11.6
2.G.A.2	Partition a rectangle into rows and columns of same-size squares and count to find the total number of them.	Lesson 11.7
2.G.A.3	Partition circles and rectangles into two, three, or four equal shares, describe the shares using the words *halves, thirds, half of, a third of,* etc., and describe the whole as two halves, three thirds, four fourths. Recognize that equal shares of identical wholes need not have the same shape.	Lessons 11.8, 11.9, 11.10, 11.11

Index

A

Act It Out strategy, 217–220, 503–506
Activities
Games. See Games
Hands On Activities: 13, 19, 31, 43, 87, 237, 255, 329, 335, 403, 409, 427, 467, 473, 479, 485, 541, 547, 553, 571, 583, 589, 603, 609, 615, 627, 639, 653, 665, 671, 723, 729, 735, 741, 747
Take Home Activity, 16, 22, 28, 34, 39, 46, 52, 58, 64, 78, 84, 90, 96, 102, 108, 113, 120, 126, 132, 138, 144, 166, 172, 178, 184, 190, 195, 202, 208, 214, 220, 226, 240, 246, 252, 258, 264, 270, 275, 282, 288, 294, 300, 306, 320, 326, 332, 338, 344, 349, 356, 362, 368, 374, 380, 394, 400, 406, 412, 417, 424, 430, 436, 442, 448, 470, 476, 482, 488, 493, 500, 506, 512, 518, 524, 530, 544, 550, 556, 562, 567, 574, 592, 606, 612, 618, 630, 636, 642, 656, 662, 667, 674, 680, 686, 708, 714, 720, 726, 732, 738, 743, 750, 756, 762, 768
Addends
adding 3 one-digit, 181–184
adding 3 two-digit, 297–300
adding 4 two-digit, 303–306
breaking apart to add, 176–177, 237–240, 249–252, 397–400
defined, 170
missing, See Addends, unknown
order of, 169–172, 181–184, 297–300, 303–306
unknown, 183, 194, 212–214, 232, 285–287, 292–293, 300, 305, 309, 371–373, 622–623, 648
Addition
adding three 1-digit addends, 181–184
basic facts, 163–166, 169–172, 175–178, 181–184, 187–190, 211–214
basic fact strategies
doubles and near doubles facts, 163–166
make a ten, 175–178, 182–184
use related facts, 187–190
breaking apart addends to add, 176–177, 237–240, 249–252, 397–400
of equal groups, 217–220, 223–226
regrouping in, 255–258, 261–264, 267–270, 273–275, 280–282, 297–300, 303–306, 403–406, 409–412, 415–417
relate to subtraction, 187–190, 359–362
represented with bar models, 187–188, 205–208, 285–288, 377–380
three-digit numbers, 391–394, 397–400, 403–406, 409–412, 415–417
breaking apart to add, 397–400
regrouping, 403–406, 409–412, 415–417
to find differences, 187–189, 359–362
two-digit numbers, 237–240, 243–246, 249–252, 255–258, 261–264, 267–270, 273–275, 279–282, 285–288, 291–294, 297–300, 303–306
breaking apart addends to add, 237–240, 249–252
finding sums of 3 two-digit numbers, 297–300
finding sums of 4 two-digit numbers, 303–306
using compensation, 243–246
using models and quick pictures, 237–238, 243, 255–258, 261–264, 267, 329, 335–338, 341, 391–394, 397, 403, 409, 419, 421, 427

write equations to represent problems, 205–208, 211–214, 285–288, 291–294

Algorithms
alternative, 237–240, 243–246, 249–252, 317–320, 323–326, 397–400
standard, 261–264, 267–270, 273–275, 279–282, 335–338, 341–344, 347–349, 353–356, 403–406, 409–412, 415–417, 421–424, 427–430, 433–436

a.m., 527–530

Analog clocks, 509–512, 515–518, 521–524, 527–530

Analyze, 319, 326, 338, 448, 470, 719

Angles
defined, 730
in two-dimensional shapes, 729–732, 735–738

Apply, 586, 655

Assessment
Show What You Know, 10, 72, 160, 234, 314, 388, 464, 538, 600, 650, 702
Mid-Chapter Checkpoint, 40, 114, 196, 276, 350, 418, 494, 568, 624, 668, 744
Chapter Review/Test, 67–70, 147–150, 229–232, 309–312, 383–386, 451–454, 533–536, 595–598, 645–648, 689–692, 771–774

Attend to Precision, 411

Bar graphs,
defined, 672
making, 677–680, 683–686
reading, 671–674
using to solve problems, 671–674, 677–680, 683–686

Bar models
addition problems, 187–188, 205–208, 285–288, 291, 377–380
multistep problems, 377–380
subtraction problems, 205–208, 365–368, 377–380

Basic Facts, 163–166, 169–172, 175–178, 181–184, 187–190, 193–195, 199–202, 205–208, 211–214

Break Apart strategy
addition, 176–177, 237–240, 249–252, 397–400
subtraction, 317–320, 323–326

Centimeters, 603–606, 609–612, 615–618, 621–624, 627–630

Cent sign, 468–470, 473–476, 479–482, 485–488, 491, 497

Chapter Review/Test, 67–70, 147–150, 229–232, 309–312, 383–386, 451–454, 533–536, 595–598, 645–648, 689–692, 771–774

Circles, 705, 711

Clocks, 509–512, 515–518, 521–524, 527–530

Coins
cent sign, 468–470, 473–476, 479–482, 485–488, 491, 497
counting collections, 467–470, 479–482, 485–488
dimes, 467–470, 473–476, 479–482, 485–488, 491–493, 497–498, 503–506
nickels, 467–470, 473–476, 479–482, 485–488, 491–493, 497–500, 503–506
pennies, 467–470, 473–476, 479–482, 485–488, 491–493, 497–500, 503–506
quarters, 473–476, 479–482, 485–488, 492–493, 497–500

Common Core State Standards, H11–H16

Communicate math ideas. See Math Talk

Compare
defined, 142
numbers, 135–138, 141–144
numbers using symbols 141–144

Cones, 705–708, 711
Correlations, Common Core State Standards, H11–H16
Counting
by 1s, 55–58, 61, 468
by 5s, 55–58, 61–64, 468
by 10s, 55–58, 61–64, 129–132, 468
by 100s, 61–64, 129–132
coins to find total value, 467–470, 473–476, 479–482, 485–488, 493, 497–499
count on and count back by 10, 123–126
count on and count back by 100, 123–126
Counting patterns
within 100, 55–58
within 1,000, 61–64, 128–132
Cubes
building rectangular prisms from, 718–720
faces, edges, and vertices, 711–714
identify and describe, 705–708, 717
Curious About Math, 9, 71, 159, 233, 313, 387, 463, 537, 599, 649, 701
Cylinders, 705–708, 711, 717

D

Data
bar graphs
defined, 672
making, 677–680, 683–686
reading, 671–674
using to solve problems, 671–674, 677–680, 683–686
line plots, 589–592, 686
picture graphs
defined, 660
making, 665–667
reading, 659–662
using to solve problems, 659–662, 665–667, 671
surveys, 653–656
tally charts, 653–656, 659, 666–667
Decimal point
in money amounts, 492–493, 498–500, 503–506
Describe/Explain a Method, 344
Digital clocks, 510–512, 516–518, 521–524, 527–530
Digits
defined, 26
values of, 25–28, 31–34, 37–39, 87–90, 93–96, 99–102, 111–113, 123–126, 279–282, 353–356
Dimes, 467–470, 473–476, 479–482, 485–488, 491–493, 497–500, 503–506
Dollar, 491–493, 497–500, 503–506
Dollar sign, 492–493, 497–500, 503–506
Doubles and near doubles facts, 163–166
Draw a Diagram strategy, 285–288, 365–368, 565–567, 621–623, 765–768

E

Edges, 712–714
Equal groups, 20–22, 217–220, 223–226
Equal parts of a whole, 747–750, 753–756, 759–762, 765–768
Essential Question
Essential Question appears in every Student Edition lesson. Some examples are: 13, 181, 329, 553, 671, 711
Estimation
of lengths in centimeters, 609–612
of lengths in feet, 577–580
of lengths in inches, 553–556
of lengths in meters, 633–636
Evaluate, 300, 306
Even numbers
defined, 14
model and classify, 13–16
as a sum of two equal addends, 19–22
Expanded form, 31–34, 37–40, 43–46, 93–96, 111–113
Explain, 524, 550, 574, 674

Faces
defined, 712
of three-dimensional shapes, 711–714
Facts. See Basic Facts
Family Involvement
Take Home Activity, 16, 22, 28, 34, 39, 46, 52, 58, 64, 78, 84, 90, 96, 102, 108, 113, 120, 126, 132, 138, 144, 166, 172, 178, 184, 190, 195, 202, 208, 214, 220, 226, 240, 246, 252, 258, 264, 270, 275, 282, 288, 294, 300, 306, 320, 326, 332, 338, 344, 349, 356, 362, 368, 374, 380, 394, 400, 406, 412, 417, 304, 424, 430, 436, 442, 448, 328, 470, 476, 482, 488, 493, 500, 506, 512, 518, 524, 530, 392, 544, 550, 556, 562, 574, 580, 586, 592, 606, 612, 618, 630, 636, 642, 656, 662, 667, 674, 680, 686, 708, 714, 720, 726, 732, 738, 743, 750, 756, 762, 768
Feet, 571–574, 577–580
Find a Pattern strategy, 49–52
Fluency
addition and subtraction within 20, 163–166, 169–172, 175–178, 181–184, 187–190, 193–195, 199–202, 205–208, 211–214
addition and subtraction within 100, 237–240, 243–246, 249–252, 255–258, 261–264, 267–270, 273–275, 279–282, 297–300, 303–306, 317–320, 323–326, 329–332, 335–338, 341–344, 347–349, 353–356, 359–362
For the Teacher
In most Student Edition lessons. Some examples are: 13, 37, 359, 409, 633,705
Fourths, 747–750, 753–756, 759–762, 765–768
Fractions, foundation for,
equal parts of a whole, 747–750, 753–756, 759–762, 765–768

Games
5 and 10 Count, 466
Caterpillar Chase, 162
Count the Sides, 704
2-Digit Shuffle, 390
Estimating Length, 602
Fish for Digits, 74
Longer or Shorter?, 540
Making Tens, 652
Subtraction Search, 316
Three in a Row, 12
What is the Sum?, 236
Glossary, H1–H10
Go Deeper, 46, 96, 101, 126, 132, 166, 172, 178, 202, 207, 214, 226, 240, 270, 288, 306, 325, 331, 349, 356, 368, 373, 380, 400, 412, 442, 482, 500, 530, 629, 655, 662, 667, 673, 680, 686, 713, 714, 720, 725, 756, 768
Graphic Organizers, 11, 73, 161, 235, 315, 389, 465, 539, 601, 651, 703
Graphs and charts
bar graphs
defined, 672
making, 677–680, 683–686
reading, 671–674
using to solve problems, 671–674, 677–680, 683–686
line plots, 589–592
picture graphs
defined, 660
making, 665–667
reading, 659–662
using to solve problems, 659–662, 665–667, 671
tally charts, 653–656, 659, 666–667

Halves, 747–750, 753–756, 759–762, 765–768
Hands On activities. See Activities

Hands On lessons, 13–16, 87–90, 485–488, 541–544, 547–550, 559–562, 571–574, 603–606, 615–618, 627–630, 639–642, 741–743
Hexagons, 723–726, 730–731
Home Activity. See Family Involvement; Take Home Activity
Hundred chart, 10, 55
Hundreds
  counting patterns with, 61–64, 123–126, 130–132
  defined, 76
  grouping tens as, 75–78
  place value, 75–78, 81–84, 87–90, 93–96, 99–102, 105–108, 111–113, 117–120, 141–144

I

Inches, 541–543, 547–550, 553–556, 559–562, 565–567, 571–574, 589–592
Inverse relationship
  of addition and subtraction, 187–190, 194
  between size of units and number of units needed to measure, 571–574, 627–630

K

Key, used in a picture graph, 659–662, 665–667, 671

Length
  add and subtract, 565–567, 621–623
  choosing tools for measuring, 583–586
  comparing, 639–642
  data displayed in line plots, 589–592, 686
  estimation of, 553–556, 577–580, 609–612, 633–636
  in feet, 571–574, 577–580
  in inches, 541–544, 547–550, 553, 559–562, 565–567, 571–574
  in meters, 627–630, 633–636
  inverse relationship between size of units and number of units needed to measure, 571–574, 627–630
Line plots, 589–592, 686
Listen and Draw
  Listen and Draw appears in most Student Edition lessons. Some examples are: 25, 49, 391, 415, 671, 753
  Math Story, 1–8, 151–158, 455–462, 693–700
Look for Structure, 52, 183, 720

Make a Model strategy, 135–138, 421–424
Make Arguments, 15, 119, 435
Make Connections, 34, 46, 189, 220, 246, 374, 476, 512, 708, 714, 738, 750
Make Sense of Problems, 28, 252, 258, 264, 288, 294, 424, 430, 488, 506
Manipulatives and Materials
  base-ten blocks, 31, 43, 87–90, 135–138, 237, 255, 261, 335, 403, 409, 421–424, 427
  centimeter rulers, 615–618, 628–630, 639–642
  color tiles, 541–543, 548, 741–743
  connecting cubes, 13–14, 19, 653, 665
  counters (two-color), 217–220, 223
  inch rulers, 560–562, 572–574, 577–580, 584–585, 589–592
  measuring tape, 584–586
  pattern blocks, 735, 747
  play money 467, 473, 479, 485–488, 503–506
  unit cubes 603–606, 615, 717–720
  yardstick, 584–585
Mathematical Practices
  MP1 Make sense of problems and persevere in solving them 49, 193, 205, 211, 217, 273, 285, 291, 365, 371, 377, 421, 503, 565, 621, 659, 665, 671, 683, 765

MP2 Reason abstractly and quantitatively 135, 187, 211, 285, 291, 365, 371, 377, 389, 565, 621, 639, 659, 671
MP3 Construct viable arguments and critique the reasoning of others 13, 19, 55, 117, 267, 347, 371, 677, 683
MP4 Model with mathematics 31, 49, 87, 135, 205, 211, 223, 237, 243, 261, 285, 291, 335, 365, 371, 377, 421, 479, 485, 491, 497, 503, 565, 589, 621, 639, 653, 665, 677, 683, 717, 723, 729, 735, 759, 765
MP5 Use appropriate tools strategically 199, 217, 255, 317, 323, 329, 359, 391, 541, 547, 559, 571, 583, 589, 603, 615, 627, 711, 741, 753
MP6 Attend to precision 25, 37, 43, 75, 99, 117, 141, 181, 223, 237, 243, 249, 267, 279, 297, 303, 341, 353, 391, 397, 403, 409, 415, 427, 439, 445, 467, 473, 509, 517, 521, 527, 389, 541, 547, 553, 559, 571, 577, 589, 603, 609, 615, 627, 633, 654, 659, 665, 671, 677, 705, 711, 729, 735, 747, 753, 759, 765
MP7 Look for and make use of structure 3, 19, 43, 49, 55, 61, 75, 81, 87, 93, 99, 105, 111, 117, 123, 129, 163, 169, 175, 187, 217, 255, 261, 273, 279, 329, 347, 353, 467, 473, 491, 497, 503, 527, 389, 553, 571, 577, 609, 627, 633, 723, 729
MP8 Look for and express regularity in repeated reasoning 19, 43, 75, 81, 93, 141, 169, 175, 181, 187, 199, 249, 267, 297, 303, 317, 323, 341, 359, 397, 403, 409, 415, 301, 427, 433, 439, 445, 479, 485, 509, 515, 521, 541, 583, 603, 741, 747

Math on the Spot, 16, 21, 27, 33, 39, 45, 51, 63, 77, 83, 95, 120, 125, 144, 165, 166, 171, 177, 184, 190, 195, 201, 207, 213, 219, 226, 239, 245, 251, 257, 258, 263, 269, 275, 281, 287, 293, 299, 305, 319, 325, 332, 338, 343, 349, 355, 361, 367, 373, 380, 393, 399, 405, 412, 417, 303, 423, 429, 436, 441, 447, 448, 327, 469, 470, 475, 487, 493, 499, 505, 511, 518, 524, 529, 391, 544, 550, 555, 562, 567, 574, 580, 586, 592, 606, 612, 618, 623, 630, 636, 641, 642, 656, 661, 667, 673, 679, 685, 686, 707, 713, 719, 726, 731, 737, 743, 749, 750, 756, 762, 767

Math Story, 1–8, 151–158, 455–461, 693–700

Math Talk
Math Talk appears in every Student Edition lesson. Some examples are: 13, 163, 341, 553, 659, 735

Measurement. See Length

Meters, 627–630, 633–636

Mid-Chapter Checkpoint, 40, 114, 196, 276, 350, 418, 494, 568, 624, 668, 744

Midnight, 528

Minutes, 510–512, 515–518, 521–524

Model and Draw
Model and Draw appears in most Student Edition lessons. Some examples are: 14, 164, 342, 542, 660, 736

Money
cent sign, 467–470, 473–476, 479–482, 485-488, 491, 497
counting collections, 467–470, 473–476, 479–482
dimes, 467–470, 473–476, 479–482, 485-488, 491–493, 497–500, 503–506
dollar, 491–493, 497–500, 503–506
dollar sign, 492–493, 497–500, 503–506
equivalencies, 485–488, 491–493
nickels, 467–470, 473–476, 479–482, 485-488, 491–493, 497–500, 503–506
pennies, 467–470, 473–476, 479–482, 485-488, 492–493, 497–500
quarters, 473–476, 479–482, 485–488, 492–493, 497–500

Multistep problems, 126, 131, 184, 190, 214, 288, 293, 319, 377–380, 405, 412, 436, 679, 680

N

Nickels, 467–470, 473–476, 479–482, 485–488, 491–493, 497–500, 503–506

Noon, 528, 530

Number lines, 199–201, 318–319, 323–325, 360–362, 374, 565–567, 621–623

Number patterns, 49–52, 55–58, 61–64, 129–132, 195

Numbers
- classifying as even or odd, 13–16
- comparing, 135–138, 141–144
- different forms of, 111–113
- different ways to write, 37–40, 105–108, 117–120
- expanded form, 31–34, 37–39, 93–96, 111–113
- in patterns, 49–52, 55–58, 61–64, 129–132, 195
- place value and, 25–28, 81–84, 87–90, 93–96, 99–102, 105–108, 111–113, 117–120, 123–126,
- representing in different ways, 43–46, 117–120

word form, 38–39, 105–108, 111–113

Odd numbers, 13–16, 19, 21

On Your Own
- On Your Own appears in every Student Edition lesson. Some examples are: 15, 183, 349, 549, 673, 737

Order of addends, 169–172, 181–184, 297–300, 303–306

Partitioning shapes, 741–743, 747–750, 753–756, 759–762, 765–768

Patterns. *See* Counting Patterns

Pennies, 467–470, 473–476, 479–482, 485–488, 491–493, 497–500, 503–506

Pentagons, 724–726, 730–732, 736, 738

Personal Math Trainer, 10, 22, 52, 72, 102, 144, 160, 172, 214, 234, 252, 264, 314, 332, 344, 388, 394, 430, 436, 464, 500, 538, 544, 586, 642, 650, 680, 686, 702, 720, 750

Picture graphs, 660
- making, 665–667
- reading, 659–662
- using to solve problems, 659–662, 665–667, 671

Place value,
- comparing numbers using, 141–144
- and counting patterns, 123–126, 129–132
- 2-digit numbers, 25–28, 31–34, 37–39, 43–46
- 3-digit numbers, 75–78, 81–84, 87–90, 93–96, 99–102, 105–108, 111–113, 117–120, 123–126
- in estimation, 141–144

p.m., 527–530

Problem Solving
- Lessons, 49–52, 135–138, 217–220, 285–288, 365–368, 421–424, 503–506, 565–568, 621–624, 683–686, 765–768
- Multistep problems, 126, 131, 184, 190, 214, 288, 293, 319, 377–380, 405, 412, 436, 679, 680
- Real World Problem Solving, 22, 34, 46, 49, 58, 64, 78, 84, 96, 102, 108, 120, 126, 132, 135, 144, 166, 172, 178, 184, 208, 214, 226, 240, 246, 252, 258, 264, 270, 282, 294, 300, 306, 320, 326, 332, 338, 344, 356, 362, 374, 380, 394, 400, 406, 412, 304, 430, 436, 442, 448, 328, 470, 476, 482, 488, 500, 512, 518, 524, 530, 392, 544, 550, 556, 562, 574, 580,586, 592, 606, 612, 618, 630, 636, 642, 656, 662, 674, 680, 686, 708, 714, 720, 726, 732, 756
- Unlock the Problem, 49, 135, 217, 285, 365, 421, 503, 565, 621, 683, 765

See also Problem Types, for word problems

Problem Solving Strategies
act it out, 217–220, 503–506
draw a diagram, 285–288, 365–368, 565–567, 621–623, 765–768
find a pattern, 49–52
make a model, 135–138, 421–424
make a graph, 683–686

Problem Types, for word problems
Add to
Change unknown, 285, 378
Result unknown, 153, 154, 163, 172, 178, 184, 214, 230, 237, 243, 252, 264, 270, 279, 286–288, 291, 300, 306, 310, 320, 326, 359, 368, 377, 379–380, 391, 623, 674
Start unknown, 213, 293
Compare
Bigger unknown, 343, 642
Difference unknown, 190, 206–208, 214, 231, 232, 326, 338, 356, 366–367, 372, 384–386, 412, 423, 439, 648, 654–656, 659–662, 672, 674, 680
Smaller unknown, 319, 332, 367, 447, 612
Put Together/Take Apart
Addend unknown, 155, 184, 208, 212–213, 232, 252, 286–287, 292, 332, 347, 367, 374, 385–386, 422, 430, 433, 442, 448, 692
Both Addends unknown, 246, 264, 282, 311, 380, 448, 606, 662, 768
Total unknown, 152, 156, 166, 169, 172, 175, 178, 184, 187, 190, 193, 196, 205, 207–208, 212–214, 230, 233, 237, 240, 246, 252, 255, 258, 261, 264, 267, 273–274, 276, 282, 287–288, 292–294, 297, 300, 303, 306, 310–312, 326, 356, 377, 379, 385, 387, 394, 400, 403, 406, 409, 412, 415, 418, 434–436, 452, 566–567, 595, 598, 622, 636, 655–656, 661–662, 671–674, 677, 680, 692
Take from
Change unknown, 187, 213, 320, 326, 371, 373, 427, 430
Result unknown, 159, 187, 190, 199, 205–207, 212, 229, 313, 320, 323, 329, 332, 335, 338, 341, 344, 348, 350, 353, 359, 362, 365, 367–368, 372, 379, 384–386, 394, 421, 423–424, 430, 433, 436, 445–446, 453, 565–567, 621–623, 642
Start unknown, 338, 373, 424

Properties of Addition,
add in any order, 169–172
adding zero, 170–171
grouping addends in different ways, 181–184, 297–300, 303–306

Q

Quadrilaterals, 723–726, 730–732, 735–738, 741–743

Quarters, 473–476, 479–482, 485–488, 492–493, 497–500

Quick Pictures, 27, 32–33, 49, 87–90, 93, 99, 112–113, 117, 123, 135–137, 141, 238, 243, 255–258, 261–264, 267, 309–310, 329, 335–338, 341, 391–394 397, 403, 409, 415, 427, 433.

R

Real World
Listen and Draw, 25, 31, 37, 43, 75, 81, 87, 93, 99, 111, 117, 123, 129, 141, 163, 169, 175, 187, 193, 199, 205, 211, 223, 237, 243, 255, 261, 267, 273, 279, 291, 297, 303, 323, 329, 335, 341, 347, 353, 359, 371, 377, 391, 403, 409, 415, 427, 433, 439, 445, 467, 473, 479, 485, 491, 497, 509, 515, 521, 527, 541, 547, 553, 559, 571, 583, 589, 603, 609, 615, 627, 633, 639, 659, 671, 677, 705, 717

Problem Solving, 22, 34, 46, 49, 58, 64, 78, 84, 96, 102, 108, 120, 126, 132, 97, 144, 166, 172, 178, 184, 190, 208, 214, 226, 240, 246, 252, 258, 264, 270, 282, 294, 300, 306, 320, 326, 332, 338, 344, 356, 362, 374, 380, 394, 400, 406, 412, 430, 436, 442, 448, 470, 476, 482, 488, 500, 512, 518, 524, 392, 544, 550, 556, 562, 574, 580, 586, 592, 606, 612, 618, 630, 636, 642, 656, 662, 674, 680, 686, 708, 714, 720, 726, 732, 756
Unlock the Problem, 49, 135, 217, 285, 365, 421, 503, 565, 621, 683, 765

Rectangles, 729, 731, 736–738
equal parts of, 748–750, 753–756, 759–762, 765–768
partition into rows and columns, 741–743

Rectangular prisms
faces, edges, and vertices, 711–714
identify and describe, 705–708, 717

Regrouping,
in addition, 255–258, 261–264, 267–270, 273–275, 280–282, 297–300, 303–306, 403–406, 409–412, 415–417
in subtraction, 329–332, 335–338, 341–344, 347–349, 353–356, 427–430, 433–436, 439–442, 445–448

S

Shapes
three-dimensional
attributes of, 711–714
identify and describe, 705–708
two-dimensional
angles in, 729–732, 736–738
attributes of, 723–726, 729–732, 735–738
identify and describe, 723–726
partitioned into equal parts, 741–743, 747–750, 753–756, 759–762, 765–768

Share and Show
Share and Show appears in every Student Edition lesson. Some examples are: 14, 170, 330, 542, 654, 736

Show What You Know, 10, 72, 160, 234, 314, 388, 464, 538, 600, 650, 702

Sides, 723–726, 637–732, 735–738

Skip counting. See Counting

Spheres, 705–711, 711, 717

Squares, 711–714, 723–726, 735–738

Subtraction
basic facts, 187–190, 193–195, 199–202, 205–208, 211–214
break apart strategy, 317–320, 323–326
relate to addition, 187–190
represented with bar models, 187–188, 205–207, 211, 365–368, 377–380
three-digit numbers, 421–424, 427–430, 433–436, 439–442, 445–448
regroup both hundreds and tens, 439–442
regroup hundreds, 433–436
regroup tens, 427–430
regrouping with zeros, 445–448
two-digit numbers, 317–320, 323–326, 329–332, 335–338, 341–344, 347–349, 353–356, 359–362, 365–368, 371–374, 377–380
using models and quick pictures, 329–332, 335–338, 341, 421–424, 427, 433, 439
using number lines, 199–202, 318–320, 323–326, 359–362
write equations to represent problems, 205–208, 211–214, 359, 365–368, 371–374
See also Problem Solving; Problem Types, for word problems

Surveys, 653–656

Take Home Activity, 16, 22, 28, 34, 39, 46, 52, 58, 64, 78, 84, 90, 96, 102, 108, 113, 120, 126, 132, 138, 144, 166, 172, 178, 184, 190, 195, 202, 208, 214, 220, 226, 240, 246, 252, 258, 264, 270, 275, 282, 288, 294, 300, 306, 320, 326, 332, 338, 344, 349, 356, 362, 368, 374, 380, 394, 400, 406, 412, 417, 424, 430, 436, 442, 328, 470, 476, 482, 488, 493, 500, 506, 512, 518, 524, 530, 544, 550, 556, 562, 574, 580, 586, 592, 606, 612, 618, 630, 636, 642, 656, 662, 667, 674, 680, 686, 708, 714, 720, 726, 732, 738, 743, 750, 756, 762, 768

Tally charts, 653–656, 659, 666–667

Test Prep

Chapter Review/Test, 67–70, 147–150, 229–232, 309–312, 383–386, 451–454, 533–536, 595–598, 645–648, 689–692, 771–774

Mid-Chapter Checkpoint, 40, 114, 196, 276, 350, 418, 494, 568, 624, 668, 744

ThinkSmarter, 16, 21, 22, 27, 28, 33, 34, 39, 40, 45, 46, 51, 52, 57, 58, 63, 64, 77, 78, 83, 84, 90, 95, 96, 102, 107, 108, 113, 120, 125, 126, 131, 132, 138, 144, 165, 166, 171, 172, 177, 178, 184, 190, 195, 201, 202, 207, 213, 214, 219, 220, 226, 239, 240, 245, 246, 251, 257, 258, 263, 264, 269, 270, 275, 276, 281, 282, 287, 288, 293, 294, 299, 300, 305, 306, 319, 320, 325, 326, 332, 338, 343, 344, 349, 350, 355, 356, 361, 362, 367, 368, 373, 374, 380, 393, 394, 399, 400, 405, 406, 412, 417, 423, 424, 429, 430, 436, 441, 442, 327, 328, 469, 470, 475, 476, 482, 487, 488, 493, 499, 500, 505, 506, 511, 512, 518, 524, 529, 530, 544, 550, 555, 556, 562, 567, 574, 580, 586, 592, 606, 612, 618, 623, 630, 636, 641, 642, 656, 661, 662, 667, 673, 674, 679, 680, 685, 686, 707, 708, 713, 714, 719, 720, 726, 731, 732, 737, 738, 743, 744, 749, 750, 756, 762, 767, 768

ThinkSmarter +, 22, 52, 102, 144, 172, 1256, 252, 264, 332, 344, 394, 430, 500, 530, 544, 586, 606, 642, 680, 686, 720, 750

Thirds, 747–750, 753–756, 759–762, 561–768

Thousand, 99–102

Three-digit numbers

addition, 391–394, 397–400, 403–406, 409–412, 415–417

comparing, 135–138, 141–144

composing and decomposing, 117–120

counting patterns with, 61–64, 123–126, 129–132

different forms of, 111–113

expanded form, 93–96, 111–113

place value, 81–84, 87–90, 93–96, 99–102, 111–113, 123–126, 141–144

subtraction, 421–424, 427–430, 433–436, 439–442, 445–448

using quick pictures to represent, 87–90, 93, 99, 117, 123, 391–394, 397, 403, 409, 415, 421–424, 427, 433, 439

word form, 105–108, 111–113

Three-dimensional shapes

attributes of, 711–714

build, 717–720

identify and describe, 705–708

Time

a.m. and p.m., 527–530

clocks

analog, 509–512, 515–518, 521–524, 527–530

digital, 510–512, 516–518, 521–523, 527–529

noon and midnight, 528

telling time, 509–512, 515–518, 521–524, 527–530

Triangles, 723–726, 729–732, 735–738, 551–750, 753

Try Another Problem, 50, 136, 218, 286, 366, 422, 504, 566, 622, 684, 766

Two-digit numbers
addition, 237–240, 243–246, 249–252, 255–258, 261–264, 267–270, 273–275, 279–282, 285–288, 291–294, 297–300, 303–306
composing and decomposing, 43–46, 49–52
counting patterns with, 55–58
different ways to represent, 31–34, 37–39, 43–46, 49–52
expanded form, 31–34
place value, 25–28, 31–34, 37–40, 43–46
subtraction, 317–320, 323–326, 329–332, 335–338, 341–344, 347–349, 353–356, 359–362, 365–368, 371–374, 377–380
word form, 37–39

Two-dimensional shapes
angles in, 729–732, 736–738
attributes of, 723–726, 729–732, 735–738
partitioned into equal parts, 747–750, 753, 756, 759–762, 765–768

Units of Measure. See Length
Unlock the Problem, 49, 135, 217, 285, 365, 421, 503, 565, 621, 683, 765
Use Diagrams, 726, 732, 768
Use Graphs, 661
Use Models, 470, 517
Use Reasoning, 22, 143, 213, 327, 544

Venn diagrams, 102, 161
Verify the Reasoning of Others, 270
Vertex/Vertices, 711–714, 723–726
Vocabulary
Chapter Vocabulary Cards, At the beginning of every chapter.
Vocabulary Builder, 11, 73, 161, 235, 315, 389, 465, 539, 601, 651, 703
Vocabulary Game, 12A, 74A, 162A, 236A, 316A, 390A, 466A, 540A, 602A, 652A, 704A
Vocabulary Review, 11, 73, 157, 161, 235, 315, 389, 461, 465, 539, 601, 651, 703
Vocabulary Builder, 11, 73, 161, 235, 315, 389, 465, 539, 601, 651, 703

Word form of numbers, 37–39, 105–108, 111–113
Word Problems. See Problem Solving, Problem Types
Write Math, 16, 22, 28, 34, 46, 58, 64, 78, 84, 90, 96, 102, 108, 120, 126, 132, 138, 144, 157, 166, 172, 178, 184, 190, 202, 208, 214, 220, 226, 240, 246, 252, 258, 264, 270, 275, 282, 288, 294, 300, 306, 320, 326, 332, 338, 344, 356, 362, 368, 374, 380, 394, 400, 406, 412, 424, 430, 436, 442, 448, 461, 470, 476, 482, 488, 500, 506, 512, 518, 524, 530, 544, 550, 556, 562, 574, 580, 586, 592, 606, 612, 618, 630, 636, 642, 656, 662, 674, 680, 686, 708, 714, 726, 732, 738, 756, 762, 768

Yardstick, 583–586